Homes in Space

Graham Rickard

Lerner Publications Company
Minneapolis

All words printed in **bold** are explained in the glossary on page 30.

Cover illustration *These astronauts are on a space walk outside the space shuttle.*

First published in the U.S. in 1989 by
Lerner Publications Company.
 First published 1988 by Wayland (Publishers) Ltd.

Library of Congress Cataloging-in-Publication Data

Rickard, Graham.
Homes in space.

Includes index.
Summary: Describes how people live in space now and may in the future.
1. Space colonies—Juvenile literature. 2. Space stations—Juvenile literature. [1. Space colonies. 2. Space stations] I. Title.
TL795.7.R52 1989 910′.0919 88-8905
ISBN 0-8225-2125-3 (lib.bdg.)

Printed in Italy by G. Canale & C. S.p.A. - Turin
Bound in the United States of America

1 2 3 4 5 6 7 8 9 10 98 97 96 95 94 93 92 91 90 89

Contents

Exploring space

Space is the name we give to the entire universe beyond the Earth's **atmosphere**. The universe is so big that it is difficult to imagine its size. It contains millions of **galaxies**, which contain millions of stars and planets, but between the stars and planets it is cold, dark, and empty. It is very difficult, dangerous, and expensive for humans to try to explore space.

People have always wanted to explore the unknown, to discover what lies beyond the limits of their knowledge and understanding. Over the centuries, explorers have discovered new continents and new places for people to live and work. We have now explored most of our planet, including its most hazardous regions, and have just begun to explore beyond it. Our fascination with discovering the unknown is shown in the many stories, films, and television shows about space travel. Over the last 30 years, the dream of entering space has started to become a reality.

There are practical reasons for going into space. Many modern materials are much easier to produce in space, and we already use many products, such as plastic railway tank cars and fuel cell power for submarines, which were developed from research done for space travel. Space travel has helped scientists gain new information. For example, **astronomers** can now use telescopes in space to see farther

Left *Carried on top of a Saturn rocket,* Skylab *blasts off in 1973.*

Left In 1965, Edward H. White was the first U.S. astronaut to leave his spacecraft while in orbit.

Below Soviet cosmonauts Alexei Leonov and Valery Kubasov are about to be launched in their* Soyuz 19 *spacecraft.

into the universe. **Satellites** are now used to relay messages all over the world. They give us valuable information about the weather and the Earth's natural resources.

The space age is just beginning. People have not yet tried to live in space for long periods of time. But one day it is likely people will live and work on permanent space stations. In the future, large space colonies could solve the problems of overcrowding on Earth. People may even travel beyond our **Solar System** to explore the distant stars and planets in outer space.

Humans and space

Over many thousands of years, human beings have adapted to living on Earth. To survive, we all need air, food, water, warmth, and light, which are all provided by our natural environment. Without the pressure of the air all around us, our blood would boil and our bodies explode. Without **gravity**, we would float around helplessly with no sense of "up" or "down." We also need clothes, a toilet, places to wash and cook, and a place to relax and sleep.

Our world provides most of us with these essential needs, but when we travel into space, we must take all our life-support systems with us. Because of this, piloted spacecraft are very large and heavy. They need enormous **rockets** to push them away from the Earth's gravity and atmosphere. It is much easier and cheaper to send unpiloted robot **probes** into space. But some space missions, such as repairing satellites, can be done only by a human crew of **astronauts**.

One of the main problems in space is the absence of gravity. Under weightless conditions, all the objects in a spacecraft, including the astronauts, float around. Water forms into floating globules, which makes drinking and washing very difficult. Astronauts have to find different ways to do the simplest everyday tasks, such as eating or sleeping. While working, they have to strap down their feet or wear slotted shoes that are attached to the floor. All

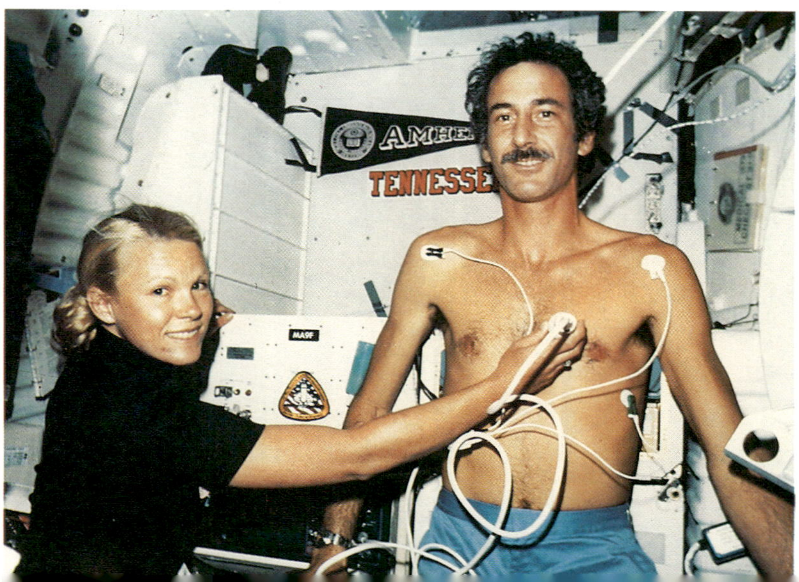

Left *Rhea Seddon checks Jeffrey Hoffman's heart rate. Each astronaut's health is monitored to see how they react to life aboard the space shuttle.*

tools and pieces of equipment must be firmly fastened in place, or they would float loose and could injure the spacecraft or the crew.

Zero gravity can cause space sickness, which, like seasickness, usually wears off after a few days. It can also cause more serious medical problems. Because they don't have to support as much body weight, muscles shrink after several months in space and bones lose calcium even more quickly. In 1987, the Soviet **cosmonaut** Yuri Romanenko lost 15 percent of his leg muscles after staying in space for a record 11 months. Instead of trying for endurance records, Soviet missions will probably be limited to six months in the future.

Above *Rhea Seddon has her feet in footholds so she won't float around the cabin while eating a meal.*

Below *An astronaut's spacesuit has a backpack that carries a life-support system.*

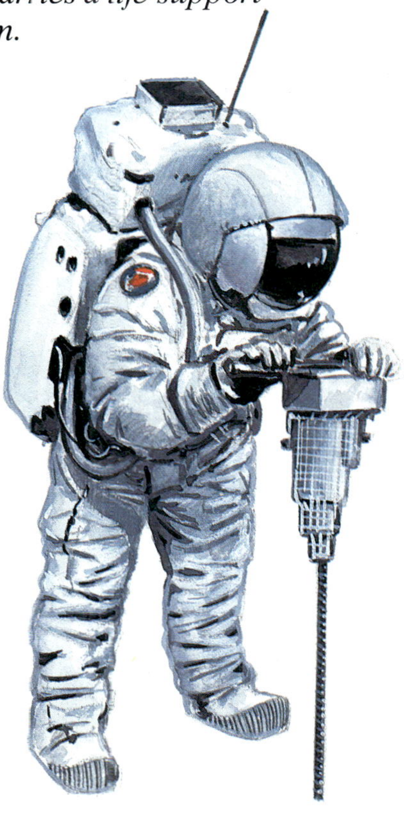

The first astronauts

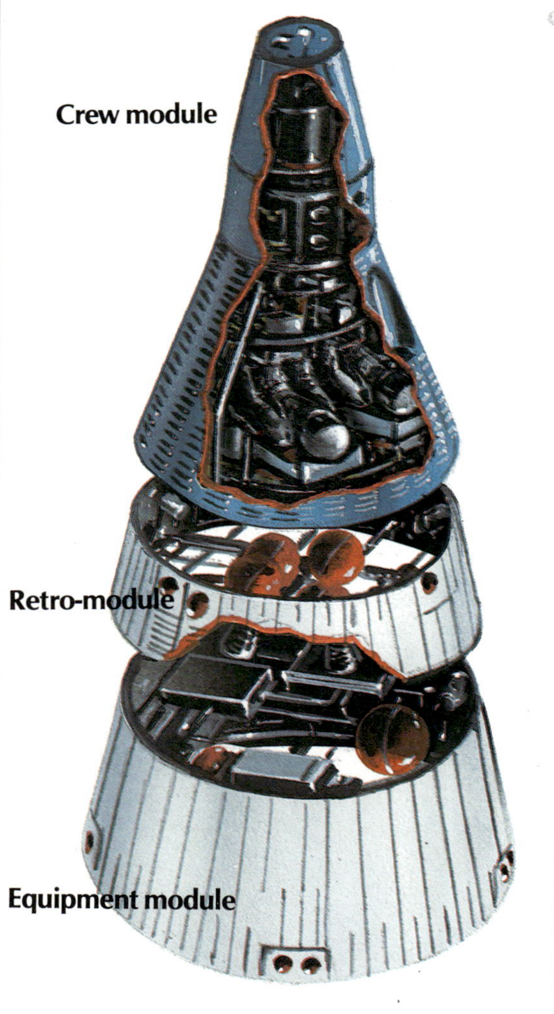

Above A drawing of a Gemini spacecraft shows the separate modules, with little space for the crew.

The space age started on October 4, 1957, when the world listened to the radio signal of the first manufactured satellite—the Soviet Union's *Sputnik 1.* In 1961, following experimental flights using dogs and monkeys, Yuri Gagarin became the first human in space. He made a complete **orbit** of the Earth before parachuting safely back to the Soviet Union. A month later, Alan Shepard became the first U.S. astronaut. He made a 15-minute flight into space but did not go into orbit. Before long, astronauts began to stay longer in space. They soon made the first space walk and the first space **rendezvous**, in which two spacecraft found each other in space.

These early astronauts were strapped into tiny **capsules** perched on top of a rocket and were unable to move from their seats. The first American spacecraft—the single-seated Mercury and the double-seated Gemini—were little bigger than telephone booths. Inside, the conditions were cramped. The astronauts' **pressurized** spacesuits provided them with enough air and warmth, but very little space. In *Gemini 7,* one astronaut was permitted to trade his spacesuit for more comfortable clothes, while the other kept his on in case of emergency. After a few days, the astronauts switched roles. By the end of the two-week flight, they had nicknamed each other Itchy and Scratchy.

Astronauts on longer missions could suck food and drink from toothpaste-like tubes, but they still had to wear heavy and cumbersome suits. Life in space was still very uncomfortable.

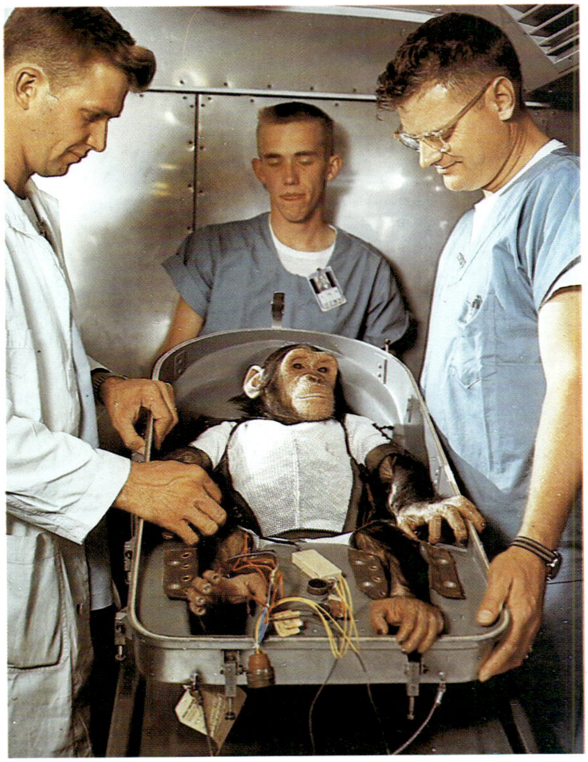

Above *Gemini 8 astronauts David Scott and Neil Armstrong wait to be lifted from the water after splashdown in the Pacific Ocean, 1966.*

Left *This chimpanzee was closely monitored on its 16-minute flight in 1961.*

The first people on the Moon

The Moon is our nearest neighbor in space. Reaching the Moon first was an obvious challenge in the "space race" of the 1960s. In the United States, the National Air and Space Administration (NASA) first sent unpiloted robot probes into orbit around the Moon to map its surface and search for possible landing sites. In 1969, the gigantic *Saturn 5* rocket was launched from Cape Kennedy in Florida. Traveling at speeds of up to 25,000 miles per hour (40,000 kilometers per hour), *Saturn 5* carried the *Apollo 11* spacecraft to the Moon in just three days. As the fuel in each stage of the rocket was used up, that stage fell away, leaving the three-part spacecraft to coast into orbit around the Moon.

The crew of three astronauts traveled in the tiny command module, *Columbia*, the only part of the spacecraft that would eventually return to Earth. *Columbia* was mounted on the service module, which carried the rocket motors and provided the astronauts with **oxygen** and electric power. The third part of *Apollo 11* was the lunar module, *Eagle*. *Eagle* separated from the rest of the spacecraft and gently dropped to the surface of the Moon, using **retro-rockets** to control its descent. Two of the astronauts were on board, and on July 20, 1969, Neil Armstrong became the first human to walk on the Moon. Armstrong and Edwin Aldrin took samples of Moon rock and set up scientific equipment before taking off in the *Eagle* to rejoin the orbiting *Columbia* for the journey back to Earth.

Left *After descending to the Moon in* Apollo 17*'s lunar module, astronaut Harrison Schmitt stands on the landing site in 1972.*

Later Apollo missions took a Lunar Roving Vehicle, or Moon buggy, to help them explore the Moon. Inside the spacecraft and on the Moon, the astronauts had to wear heavy protective suits and helmets, with a backpack which held all their life-support systems. The spacecraft were larger, but still very cramped and uncomfortable. The crew had no place to exercise, no shower or place to shave, and no place to relax.

Below *The Moon buggy was used to explore the Moon and to collect Moon rock.*

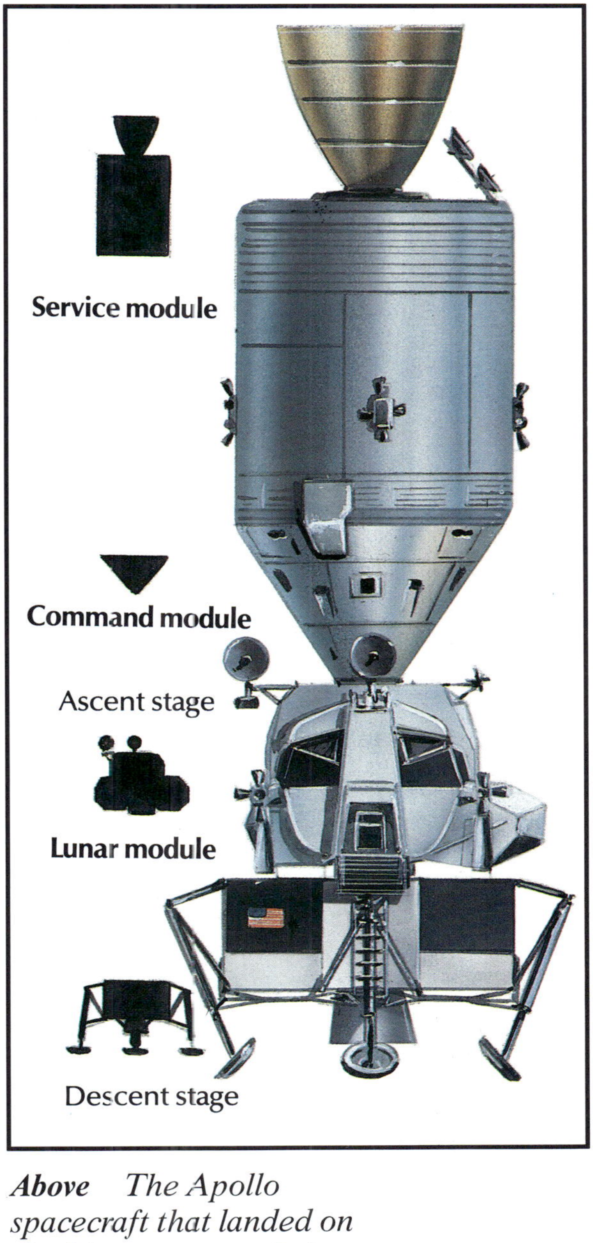

Above *The Apollo spacecraft that landed on the Moon were made in three parts.*

The first space stations

Despite the success of the Apollo program, trips into space were still fairly short. To see if astronauts could stay in space for longer periods of time, NASA needed to place a large space station in permanent orbit. NASA constructed *Skylab* from the empty shell of a used *Saturn* rocket and put it into orbit in 1973.

Skylab was large enough so that astronauts could live comfortably in space for the first time. Inside *Skylab*, astronauts could remove their spacesuits and wear ordinary clothes. There was more room for stowing spacesuits and, if air started to leak from the cabin, the astronauts could go through an air lock into another room. Two large **solar** panels supplied electrical power.

Half of *Skylab* contained an experimental chamber where the crew could test the effects of weightlessness. In the other half was a workshop for conducting experiments, a small **galley** for cooking prepacked meals, and a bathroom. There were also sleeping compartments, a shower with a vacuum to suck off the water, and a lounge.

Left Skylab *in orbit, taken from a departing Apollo spacecraft*

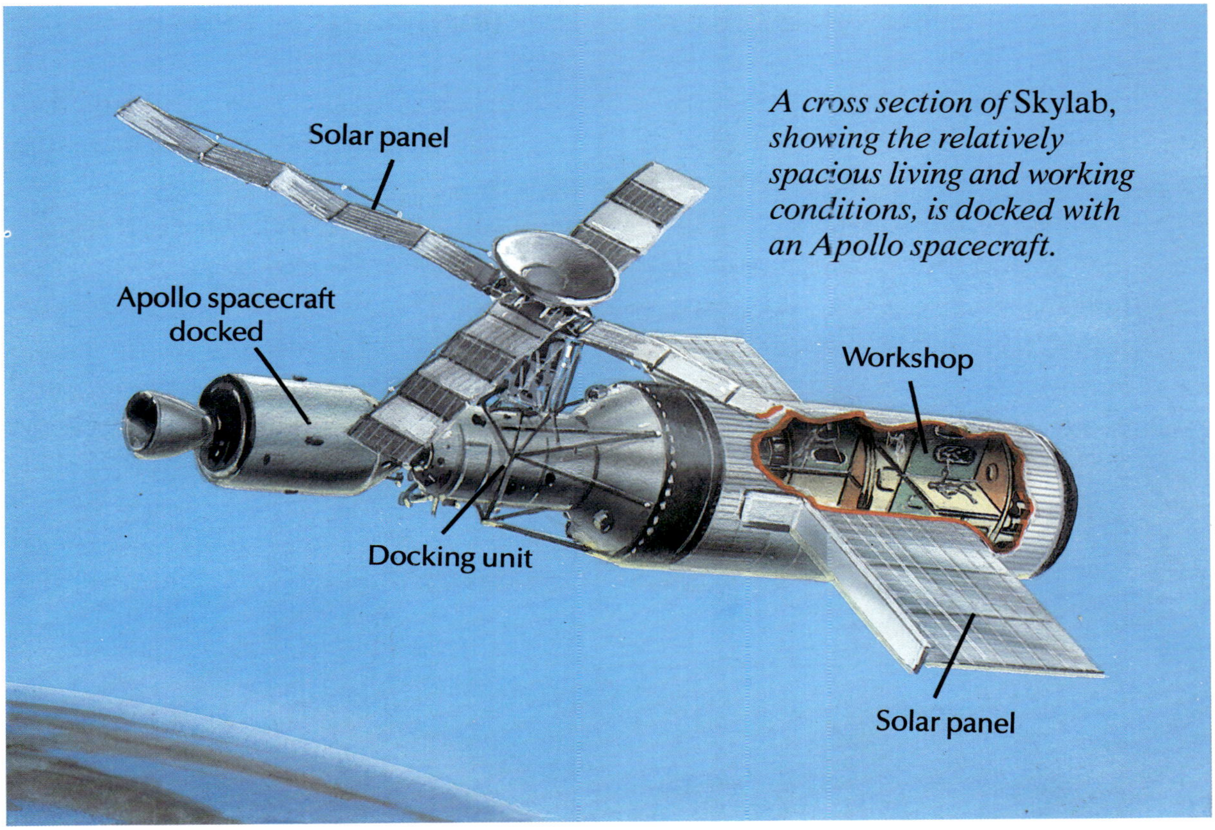

A cross section of Skylab, *showing the relatively spacious living and working conditions, is docked with an Apollo spacecraft.*

The Soviet Union launched their first *Salyut* space station in 1971, and their cosmonauts soon set records for the longest time spent in space. Both *Salyut* and *Skylab* crews experimented with growing food and manufacturing materials in space and provided scientists with valuable information. Without the force of gravity, crews needed to exercise regularly to keep in shape. Their heart rate, breathing, and blood pressure were measured to see how their bodies reacted to weightlessness.

The *Salyut* and *Skylab* crews proved that humans could live and work safely in space for long periods of time. One day, astronauts may be living in a new generation of large space stations.

The space shuttle

Space exploration is very expensive. The early rockets were very wasteful because they could be used only once. NASA started a new era in space exploration when it launched the reusable space shuttle, which takes off like a rocket and lands like an airplane. For takeoff, the shuttle is strapped to two big booster rockets and an enormous fuel tank that feeds the three rockets engines near its tail. The booster rockets and fuel tank drop away from the shuttle when the fuel is used up.

The shuttle has a large **cargo bay** to carry its **payload**. Five large computers control the navigation and life-support systems. The air is constantly filtered and recirculated at an even temperature and pressure. Fuel cells produce electricity and water for drinking and washing. And the living quarters can easily hold seven crew members.

This drawing shows a cross section of a space shuttle.

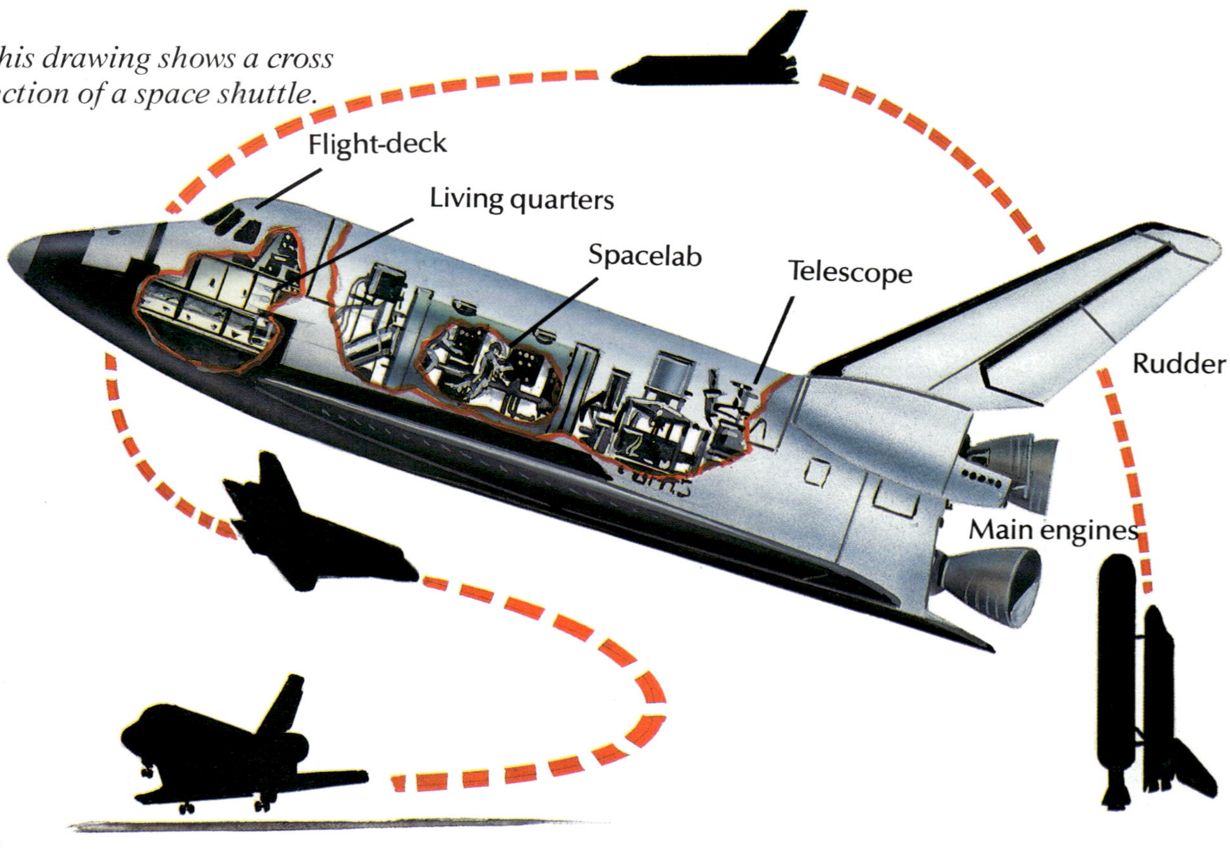

The galley contains a pantry, oven, water dispenser, and meal trays. The crew can choose meals from a variety of stored foods. To prepare freeze-dried foods, water is injected into the package, and then an astronaut squeezes the carton to mix it up before eating.

Because of zero gravity, the toilet is fitted with a seatbelt, and a hand-operated fan draws waste into a storage tank. To stop water from floating around the cabin, the sink is enclosed. The astronauts wear comfortable clothes, with plenty of pockets to keep their tools. They exercise regularly on **treadmills** and exercise bicycles to keep their bodies in shape for Earth's gravity. The sleep station is fitted with sleeping bags, which are strapped down to stop the crew from drifting around in their sleep.

The shuttle program was halted in 1986 by the tragic explosion of *Challenger*, which killed seven astronauts. Even so, the shuttle and similar spacecraft will have a very important part to play in the next stage of space exploration.

Below *The space shuttle* Enterprise

A shuttle astronaut's day

In space, there is no night or day, so astronauts still schedule their routine around a 24-hour Earth day. Typically, they have a 4-hour working day, with 9 hours for sleep and 2 for exercise. The rest of the time is spent reading, listening to music, playing in weightlessness with other astronauts, and relaxing.

After crawling out of her sleeping bag, the astronaut pushes off from the opposite wall or grabs a handrail to pull herself toward the covered sink, which has two handholes. Inside are soap and water dispensers, and a fan sucks away the dirty water into a storage tank, which is then emptied into space.

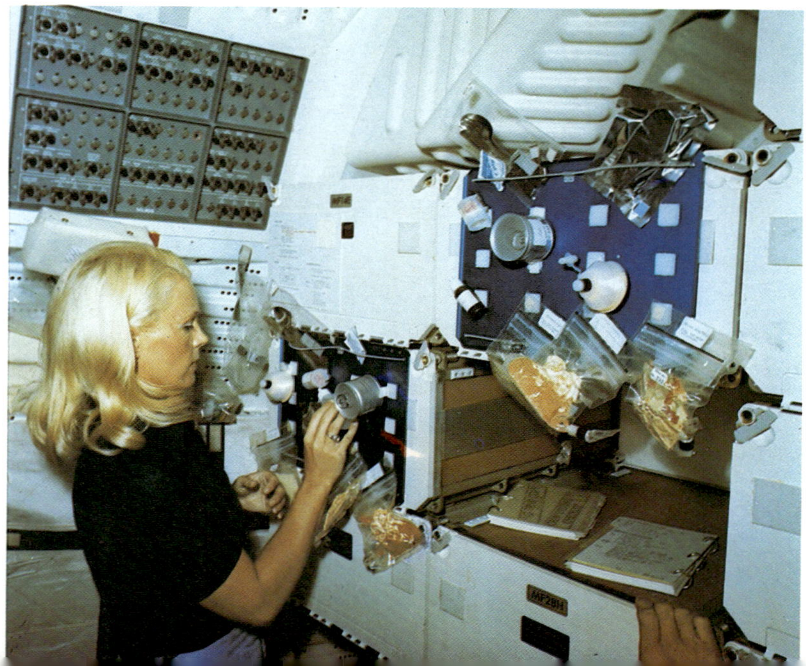

Above *An astronaut works outside the shuttle attached to a moveable arm.*

Left *Rhea Seddon assembles items on food trays inside the shuttle galley.*

After dressing in comfortable clothes, similar to a sweatsuit, the astronaut pulls herself toward the galley. In the pantry, one of the astronauts has prepared and cooked prepacked meals for the crew. The food container fits in a hole in each astronaut's tray. Floating wherever she likes, the astronaut sucks her drink from its pouch and spoons food into her mouth. The food is usually stuck together with a thick sauce to keep it from floating off the silverware and drifting around the cabin. After eating, she has time for half an hour's exercise on the treadmill before starting work.

Most work is done in the flight deck or mid deck of the space shuttle, but to launch or repair satellites and perform some of the experiments, the astronaut must go outside. Before entering the cargo bay, the astronaut must put on her bulky spacesuit and backpack, which holds all her life-support systems. She is tethered to a life-line, which stops her from drifting off into space. The thick gloves she must wear make it difficult to adjust a piece of experimental equipment or mend a faulty satellite.

Above *An astronaut exercises on a treadmill designed for conditions of zero gravity.*

Back in the cabin, the astronaut relaxes and talks to the other crew members before returning to the sleep station and strapping herself back into her sleeping bag.

Space stations

The U.S. *Skylab* and the Soviet *Salyut* and *Mir*, another Soviet space station, were fairly small experimental space stations, but they proved that astronauts can survive for almost a year in space with few ill effects.

During the next decade, shuttle-type craft may be used to build larger space stations with a permanent crew of astronauts. Shuttles would also be used to ferry crews, supplies, and equipment from Earth and to bring astronauts who have completed their long missions back to Earth.

The Soviet Union is planning to make its future space station from several Mir stations joined together. But the United States, Japan, and several countries in Europe have joined to produce a much more ambitious idea. In the 1990s, NASA plans to start building a massive and costly new space station. The main part will be a long metal frame supporting the solar panels which will provide the space station's power. At the center will be four 46-foot (14-meter) capsules, where a crew of six astronauts will live and work for up to six months at a time. The interior of the living quarters has been designed by a team of architects and will be much more comfortable and homey than previous space stations.

This new space station, *Columbus*, will consist of many different parts, including a huge satellite for observing the Earth. Information from this satellite will warn us of bad weather and tell us where to look for water, oil, and even schools of fish. *Columbus* will also contain laboratories and small factories. Perfect crystals can be grown in space for use in the electronics industry. This is because space is totally free from gravity, air pressure, dust, and pollution.

A large space station could also provide an ideal space "port" for long space journeys or for launching spacecraft. This kind of space station would be a useful workshop base for astronauts to repair and modify satellites.

Right *Within the next decade, large space stations with a permanent crew of astronauts will start a new era of space exploration.*

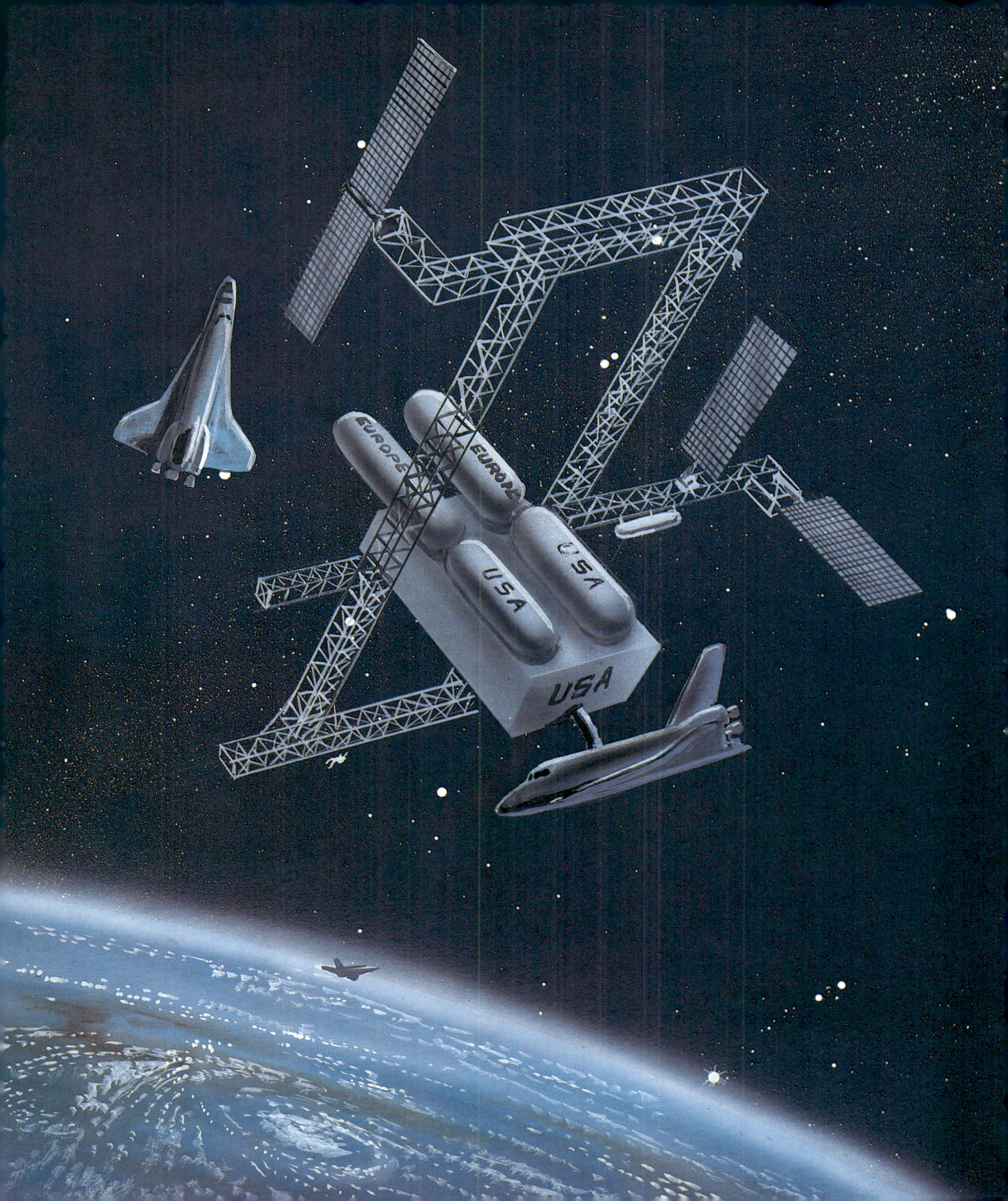
EUROPE
EUROPE
USA
USA
USA

Living on the Moon

Humans have already set foot on the Moon, which will probably be the site of the first **extraterrestrial** human **colony**. NASA hopes to have a Moon base early in the twenty-first century. Although the Moon is our nearest neighbor in space, there are many problems to be overcome before we can think of living there in permanent settlements.

Above *As the Earth rises over the Moon, it appears as a beautiful blue and white planet.*

Above *To survive the harsh conditions of the Moon, astronauts have to wear heavy spacesuits.*

The Moon has no air or water and very little gravity. As one full rotation, or day, of the Moon equals 28 Earth days, the temperature on the Moon varies from scorching hot to freezing cold. Covered with vast craters and mountains, the surface of the Moon is very rough. Unprotected by an atmosphere, the Moon is bombarded by harmful radiation from the Sun.

Despite all these disadvantages, the Moon would make a useful base for further space exploration. A colony of domeshaped buildings,

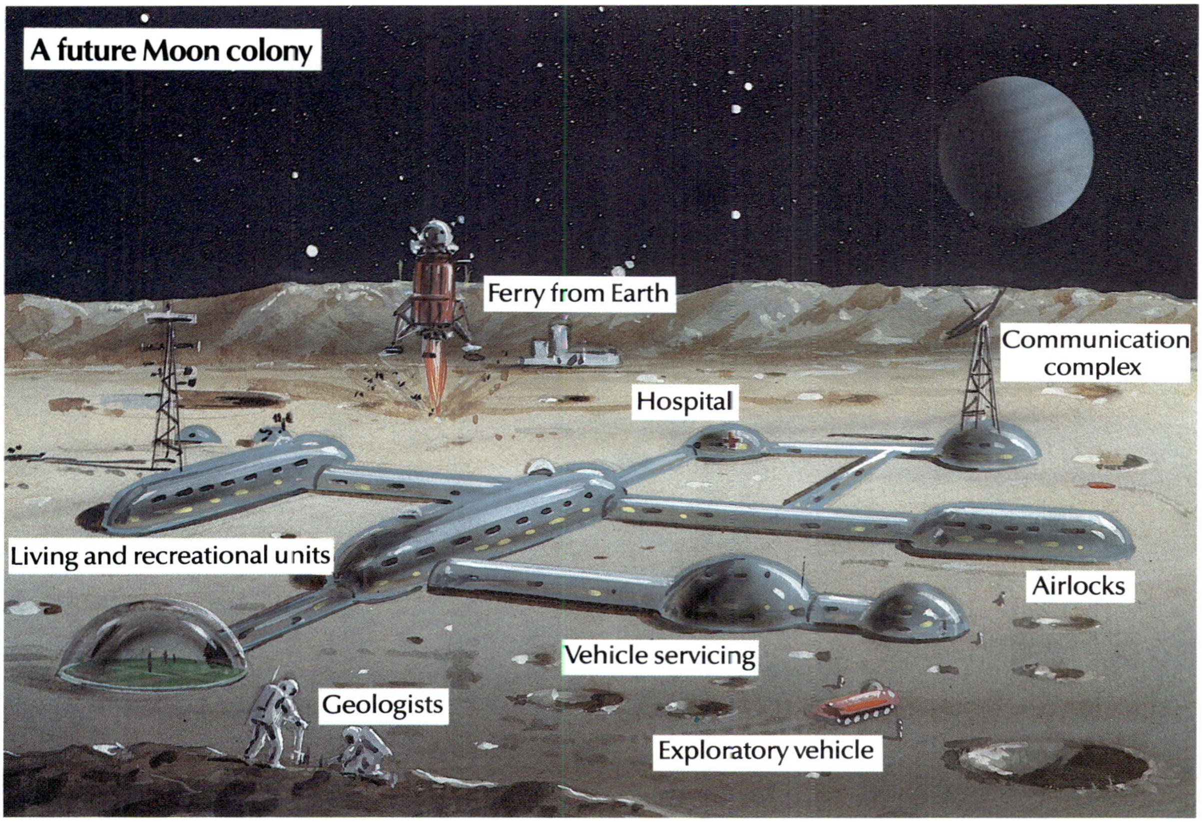

connected by tunnels, could provide a safe home for scientists and astronauts. Oxygen could be made from Moon rocks, and **hydroponic** farms could grow plants for food. Moon dwellers could stay on the Moon for years at a time. The Moon is ideal for astronomy, since there is no atmosphere or dust to block light from distant stars. Spacecraft could be built from metals mined on the Moon and, because of the low gravity, launched with very little fuel.

Scientists have recently made some troubling predictions about human life on the Moon. Because of low gravity and lack of air pressure, they believe that children who are born and raised on the Moon might grow 36 feet (11 m) tall, with bones so thin that they would not be able to survive if they ever came to Earth.

Living on the planets

Our Sun is a star, which is orbited by the Earth and the eight other planets that make up the Solar System. The problems of living on the Moon are small compared with the problems involved in setting up bases on any of the other planets. No human being has yet traveled farther than the Moon—a journey that only takes a few days. Even the nearest planets, Venus and Mars, are so far away that the journey would take up to six months in each direction. Without shuttles from Earth, the spacecraft for a trip with humans on board would need to be enormous to carry all the necessary fuel and supplies.

However, astronauts may one day visit the planets that have already been investigated by unpiloted space probes. Of all the planets, Mars seems the most friendly to human needs and is the most likely site for a future space base. After the U.S.'s Mariner probes had photographed the planet, the Viking probes landed on Mars

Below *A drawing of the planets and star that make up our Solar System.*

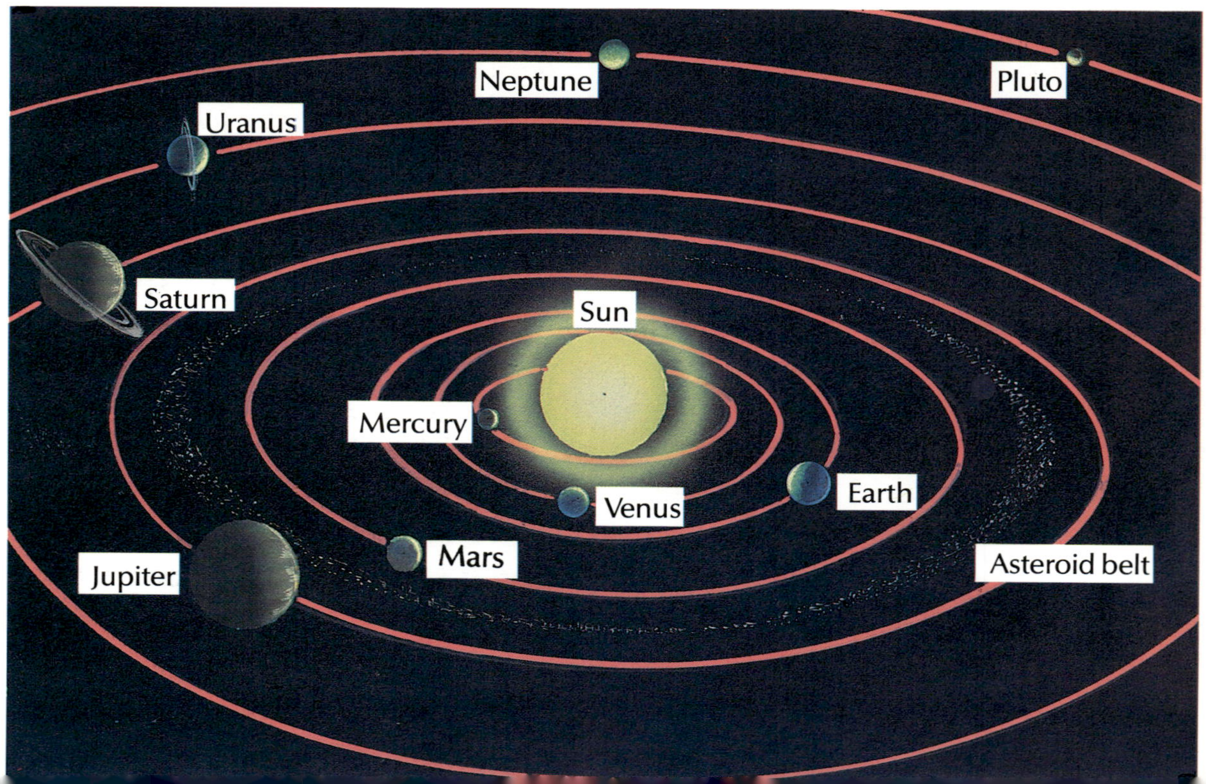

and analyzed its surface. In 1992, the U.S.S.R. plans to send a mobile robot probe to study the Martian surface. Mars does have some gravity and a thin atmosphere of **carbon dioxide**. The vast red desert that covers Mars has no water. The planet is exposed to intense bursts of radiation from the Sun, and the temperature rarely rises above freezing. But a long-term Mars mission is still thought to be possible, in spite of the dangers involved. Having arrived on Mars, a crew would have to grow their own food. The food plants could also supply an oxygen-rich atmosphere inside domes.

Above *An unpiloted Viking space probe landed on Mars in 1976 to look for signs of life.*

Above *Saturn, the second largest planet in our Solar System, is encircled by rings made of ice particles that orbit the planet.*

The other planets all pose even more serious problems for any future space traveler. Venus is very hot, with an atmosphere made up of poisonous gases. Mercury has no atmosphere at all, and its surface is covered with steep mountains and craters. Beneath its clouds, Jupiter is very hot. It consists of gases and liquids with solid rocky material only at its core. The other planets—Saturn, Neptune, Uranus, and Pluto—are all in the freezing outer parts of the Solar System and probably contain very little solid matter.

Activities in space

The space age has only just begun. The world will soon see a vast increase in space research and exploration, with many countries taking part. Already there are arguments about the kinds of activities that should be allowed in space.

It was once agreed that no weapons should be allowed in space and that everyone had an equal right to use space. But some countries, including the United States, already use military satellites to watch other countries. The U.S. is already planning weapons in space as part of its Strategic Defense Initiative (SDI) program. The U.S.S.R. is worried by the thought of being defenseless against powerful space weapons and is planning to produce similar weapons. Military bases might be set up on the Moon. Orbiting space stations could contain the very latest types of deadly weapons.

It is possible that space will continue to be a place for peaceful scientific research and international cooperation. We have already gained a great deal from space exploration, and the future rewards could be

Left *In the future, these powerful lasers could be used as space weapons.*

enormous. Many materials that are impossible to make on Earth can be produced in space because there is no gravity, air pressure, or pollution. New **alloys** are made possible in space because metals mix better when they have no weight difference. Currently, these alloys are used in oil refineries and huge aircraft used to transport helicopters, large computers, and even rockets. Research in space could benefit the field of medicine. In the near future, space laboratories and factories could provide us with many more products to improve our life on Earth.

Above *Future astronauts will have to grow their own food. These mung seedlings were grown in Spacelab.*

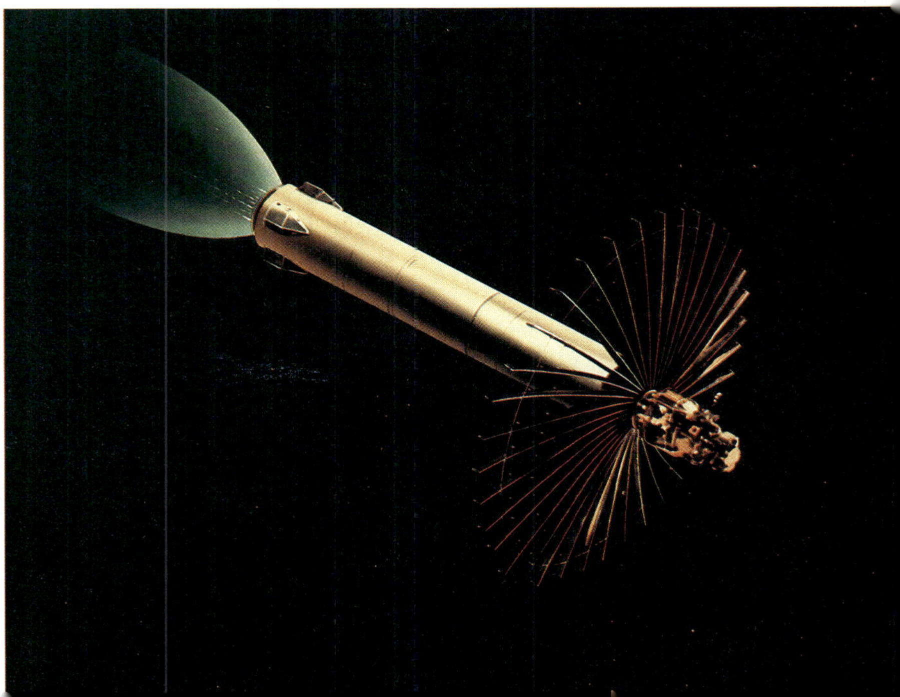

Right *An artist's impression of a weapon that could be used in the United States' SDI program to destroy missiles and satellites by colliding with them.*

Living in space colonies

Some people believe that in the future, if our world becomes even more crowded and polluted, we may have the technology that allows large groups of people to leave our planet to start a new life in a space colony. These cities in space would be contained in massive space stations high above the Earth, beyond the Moon's orbit. Such colonies would be entirely independent, with their own schools, factories, and hospitals, and with their own government. They would be in the shape of a cylinder or hollow ring, spinning in space to produce an artificial gravity by **centrifugal force**. The artificial atmosphere would be kept at a comfortable temperature and the climate pattern controlled by a computer program.

Each colony could house thousands of settlers, who would grow their own food in domes landscaped with rivers and trees. Arriving settlers would be ferried in shuttle-type vehicles to "space hotels" before traveling on to the colony. The inhabitants would use communication satellites to keep in touch with Earth and could watch vast **holograms** in space for their entertainment. They could mine **asteroids** for metals and work in factories to produce space products to trade with products from Earth. Spacecraft could be launched from the colony for deeper exploration into space. Whole generations might be able to spend their entire lives in such colonies.

At the moment, such space colonies are pure fantasy and exist only in the minds of scientists and science-fiction writers. One day they could provide the answer to many of the world's problems.

Below *Inside a space colony*

An artist's conception of a future space colony where people would live inside a giant wheel. The wheel would rotate to create artificial gravity and solar energy would be used for power and heat.

A cross section of the colony.

Traveling in outer space

The term "outer space" is used for the whole of the universe outside the Solar System. By human standards, our Solar System is enormous. Even to go to the Sun, which is 93 million miles (150 million km) away, would take more than 10 years going the speed at which we can now travel. But the Solar System is only a small part of the Milky Way galaxy, which contains thousands of millions of stars and planets. In turn, our galaxy is only a fragment of the universe. The distances in outer space are so vast that they are measured in light years. Light travels at 186,282 miles (299,792 km) per second. A light year is the distance that light can travel in one year, nearly 6 trillion miles (9.5 trillion km). Some galaxies in the universe are millions of light years away!

Traveling to the stars will be a great adventure for humankind, but there are obviously many problems to overcome. A present-day spacecraft would take thousands of years just to reach the star nearest our Sun. Many generations of astronauts would be born and die before the

Left *The Andromeda Galaxy, 2.2 million light years from Earth, is made up of millions of stars in the shape of a spiral.*

Right Pioneer 10 *flies past the Great Red Spot on Jupiter. Unpiloted probes like this pave the way for future piloted missions.*

spacecraft reached its destination. New types of engines and fuel must be found to increase the speed of space travel. Living conditions in spacecraft would have to be improved to the point where astronauts could comfortably spend their entire lives traveling in space. Some scientists have suggested that astronauts could be deep-frozen on journeys to the stars and revived automatically many years or even centuries later.

People have often wondered whether there are other inhabited planets in the universe. Future space travelers might possibly meet some form of intelligent beings living on an unknown planet in some distant galaxy.

Glossary

alloy A mixture of two or more metals
asteroid One of thousands of small planets or rocks that orbit the Sun, mainly between Jupiter and Mars
astronaut A person who travels in space
astronomer A person who studies the stars and planets
atmosphere The layers of gases surrounding a planet
capsule A small pressurized compartment for astronauts; a spacecraft
carbon dioxide A gas that we breathe out and also that plants breathe in
cargo bay The part of the space shuttle, open to space, where satellites and Spacelab are carried
centrifugal force The force that pulls an object outward from the center of rotation
colony A group of people who settle in a new place and keep ties with their original home
cosmonaut A Soviet astronaut
extraterrestrial Originating or existing outside the Earth and its atmosphere
galaxy A large group of stars and planets
galley A small kitchen
gravity The force that pulls objects toward each other
hologram A picture created by laser beams
hydroponics A way of growing plants using liquids instead of soil
orbit The path that a satellite or spacecraft follows as it revolves around a planet, or that a planet follows as it revolves around a star
oxygen A gas that we breathe in and also that plants breathe out
payload The cargo and passengers carried by a spacecraft
pressurized Filled with air under pressure, like a balloon
probe A small, unpiloted spacecraft, which is used to explore space
rendezvous The process of bringing two spacecraft together
retro-rocket A small rocket engine on a spacecraft, which is used to change the spacecraft's direction or slow it down
rocket A jet engine which moves through the air or through space by burning fuel and oxygen
satellite A natural or manufactured object that orbits a planet, a moon, or a star
solar Having to do with the Sun
Solar System The nine planets and all the moons and asteroids that revolve around the Sun
space The entire universe outside the Earth's atmosphere
treadmill A wheel with many treads which turns under the weight of a person using it for exercise
zero gravity No gravity at all, weightlessness

Books to read

From Sputnik to Space Shuttles by Franklyn M. Branley (Thomas Y. Crowell, 1986)
Flying the Space Shuttles by Don Dwiggins (Dodd, Mead and Company, 1985)
Space Challenger: The Story of Guion Bluford by Jim Haskins and Kathleen Benson (Carolrhoda Books, 1984)
Colonies in Orbit by David C. Knight (William Morrow and Company, 1977)
To Space and Back by Sally Ride and Susan Okie (Lothrop, Lee & Shepard Books, 1986)
Space Walking by Gregory Vogt (Franklin Watts, 1987)

Houses and Homes

Building Homes
Castles and Mansions
Homes in Cold Places
Homes in Hot Places
Homes in Space
Homes in the Future
Homes on Water
Mobile Homes

Picture acknowledgements

Photographs and illustrations courtesy of: pp. 4, 12 (left), 20 (left), Bruce Coleman; pp. 5 (top), 6, 7 (left), 9 (bottom), 24 (Dept. of Defense), 25 (top), 28, The Research House (NASA); pp. 5 (bottom) (Novosti), 9 (top), 10, 11 (left), 12 (right), 20 (right), 25 (bottom) (Dept. of Defense), NASA/Science Photo Library; pp. 7 (right), 8, 11 (right), 13, 14, 16, 17, 19, 21, 22, 26, 27, 29, Gordon C. Davies; p. 15, PHOTRI; p. 23, Camera Press (Ray Hamilton). Cover photograph courtesy of PHOTRI.

Index